Emergency Preparedness and Response for Offshore Oil & Gas

Table of Contents

Preface

Introduction

Prelude

Chapter 1: Understanding Offshore Oil & Gas Risks

Chapter 2: Regulatory Framework for Offshore Emergency Preparedness

Chapter 3: Emergency Response Planning

Chapter 4: Oil Spill Contingency Planning

Chapter 6: Fire and Explosion Response in Offshore Operations

Chapter 7: Managing Human Factors in Emergencies

Conclusion

Glossary

Preface

The offshore oil and gas industry operates in a highly dynamic environment, where evolving technologies, stricter regulatory requirements, and increasing expectations for safety and environmental stewardship shape the future of operations. In this ever-changing landscape, the importance of preparedness and effective response to emergencies cannot be overstated. To equip professionals with the knowledge and skills necessary to navigate these complexities, the *Gosships Learning Series* provides accessible, practical, and targeted learning resources.

Each book in this series is meticulously designed to offer foundational to intermediate knowledge with a focus on real-world application. The content aims to support industry professionals—from offshore crew members to management personnel—by offering insights that are both current and relevant to the challenges faced in offshore oil and gas operations. Coupled with certification tests, these resources ensure that learners not only gain a strong grasp of the material but also demonstrate their competency in applying this knowledge to enhance safety, compliance, and operational efficiency.

The *Gosships Learning Series* empowers professionals in the maritime and energy sectors, enabling them to stay ahead of industry changes and contribute to safer, more efficient operations. We hope that this series will enhance your professional development and open doors to greater opportunities in your career.

Introduction

Welcome to the *Gosships Learning Series*, a platform dedicated to expanding knowledge and advancing careers in the offshore oil and gas industry. This book, *Emergency Preparedness and Response for Offshore Oil & Gas*, has been carefully developed by industry experts and regulators to provide essential insights into one of the most critical aspects of offshore operations: emergency preparedness. Whether you are new to the field or looking to deepen your expertise, this resource is tailored to offer a comprehensive understanding of emergency response protocols and best practices.

In this book, we will cover the following critical areas:

- **Understanding Offshore Oil & Gas Risks**: Learn about common hazards such as blowouts, oil spills, equipment failures, and human factors, all of which can escalate into emergencies.

- **Regulatory Frameworks**: Gain insights into the international and national regulations that govern offshore emergency preparedness, ensuring safety and compliance.

- **Emergency Response Planning**: Explore the process of developing and implementing effective emergency response plans, including drills, communication protocols, and defined roles.

- **Oil Spill Contingency Measures**: Understand the key strategies for containing and responding to oil spills, including recovery techniques and regional cooperation.

- **Evacuation and Rescue Operations**: Discover best practices for planning and executing evacuation procedures, including the use of helicopters, lifeboats, and rescue vessels.

- **Fire and Explosion Response**: Delve into the specifics of fire suppression systems, firefighting techniques, and coordinating with onshore emergency teams.

- **Managing Human Factors**: Learn about the psychological and physiological responses to emergencies, and how to ensure effective leadership and decision-making under pressure.

Upon completing this book, you will be well-prepared to take an assessment that evaluates your understanding of emergency preparedness in offshore oil and gas operations. A Certificate of Achievement will be awarded to those who successfully complete the assessment, which can be obtained by visiting www.gosships.com. This certification will help validate your expertise in this critical area, allowing you to demonstrate your proficiency and preparedness to industry peers and employers.

Who Is This Book For?

This book is designed for a wide range of professionals who are involved in or aspire to work within the offshore oil and gas industry. It is an essential resource for:

- **Offshore and Maritime Personnel**: Those seeking to strengthen their understanding of emergency preparedness and improve their ability to respond effectively to offshore emergencies.

- **Shoreside Managers**: Individuals responsible for overseeing offshore operations, who need a solid understanding of industry best practices to ensure safety and compliance.

- **Aspiring Students**: Individuals looking to enter the industry with a strong foundation in emergency preparedness and response protocols.

- **Government and Regulatory Personnel**: Those tasked with enforcing safety and environmental regulations, who need to stay informed about evolving industry standards and practices.

By mastering the concepts covered in this book, you will be better equipped to handle the challenges of modern offshore operations, maintain compliance with international regulations, and contribute to a safer, more efficient work environment.

Thank you for choosing the *Gosships Learning Series* as part of your journey toward continuous learning and professional excellence. Your commitment to staying informed and prepared in this high-risk industry is crucial to ensuring the safety and sustainability of offshore oil and gas operations.

Gosships Learning Series 2024/2025

1. Hydrogen: The Fuel of the Future
2. Green Ammonia: The Next Big Thing in Shipping
3. Decarbonizing Shipping: Pathways to Zero Emissions
4. Battery Technology for Industrial Applications
5. Carbon Capture and Storage: Can It Save the Planet?
6. Biofuels 101: Turning Waste into Energy
7. Understanding LNG (Liquefied Natural Gas)
8. Methanol as a Marine Fuel
9. Offshore Wind Energy: The Future of Renewable Power
10. Tidal and Wave Energy: Harnessing the Ocean
11. Electrofuels: The Next Generation of Carbon-Neutral Fuels
12. Energy Storage Systems for Grid Reliability
13. Hydrogen Fuel Cells for Transportation
14. Solar Energy Innovations: Beyond Solar Panels
15. Smart Grids: The Backbone of Future Energy Systems
16. Ammonia-Hydrogen Blends: A Dual Fuel Solution?
17. Nuclear Power: Small Modular Reactors for a Low-Carbon Future
18. Hydropower: The Oldest Renewable Energy Source
19. Decentralized Energy Systems: Microgrids for Resilience
20. Energy Efficiency Technologies for Industry
21. Hydrogen Production from Seawater
22. Fuel Cells for Maritime Applications
23. Geothermal Energy: Unlocking Earth's Heat
24. Future of EV Charging Infrastructure
25. Synthetic Fuels: Bridging the Gap to Decarbonization
26. Cybersecurity for Maritime and Offshore Operations
27. AI and Automation in Shipping and Logistics
28. Digital Twins in Maritime: Revolutionizing Asset Management

29	Risk Management in Offshore and Maritime Operations
30	Compliance with IMO 2020 Regulations
31	Sustainable Ship Design: Reducing Environmental Impact
32	Marine Renewable Energy: Wave, Tidal, and Offshore Wind Integration
33	Ballast Water Management Systems
34	Blockchain Technology in Shipping: Improving Transpc'y & Efficiency
35	Effective Supply Chain Management for Energy Industries
36	Leadership in the Energy Transition
37	Effective Crisis Management in Maritime Operations
38	Shipyard Safety Management Systems
39	Port State Control (PSC) Inspection Readiness
40	Remote Vessel Operations and Autonomous Shipping
41	Optimizing Fleet Performance with Data Analytics
42	Maritime Environmental Regulations: Staying Ahead of Compliance
43	Advanced Maintenance Strategies: Condition Monitoring & Predictive Maintenance
44	Global LNG Market: Trends and Opportunities
45	Incident Investigation in Maritime Operations
46	International Maritime Law: Key Concepts and Applications
47	Emergency Preparedness and Response for Offshore Oil & Gas
48	Energy Transition Strategies for Oil and Gas Companies
49	Maritime Drones: Applications and Safety Considerations
50	Effective Project Management in Offshore Energy Projects

All Rights Reserved Disclaimer

The contents of this book, including but not limited to all text, graphics, images, logos, and designs, are the intellectual property of Gosships LLC and are protected by copyright law. No part of this publication may be reproduced, distributed, transmitted, displayed, or modified in any form or by any means, including photocopying, recording, or other electronic or mechanical methods, without the prior written permission of the publisher, except in the case of brief quotations in critical reviews or articles.

The information contained within this book is for educational purposes only and is provided "as is" without warranty of any kind, either expressed or implied. The authors and publishers disclaim any liability for any direct, indirect, or consequential loss or damage arising from the use of the material in this book.

For permissions or inquiries, please contact: admin@gosships.com

© 2024 Gosships LLC. All rights reserved.

Prelude

The offshore oil and gas industry operates in some of the most hazardous and remote environments in the world, often facing extreme weather, complex technical challenges, and a heightened risk of disasters such as fires, explosions, blowouts, and oil spills. Effective emergency preparedness and response are vital to minimizing the consequences of these risks, ensuring the safety of personnel, protecting the environment, and maintaining operational continuity. This expanded guide explores the key aspects of emergency preparedness and response in the offshore oil and gas industry, covering hazard identification, regulatory requirements, emergency response planning, oil spill contingency measures, evacuation and rescue operations, firefighting strategies, and the management of human factors.

Chapter 1

Understanding Offshore Oil & Gas Risks

1.1 Common Offshore Hazards

Offshore oil and gas operations face a unique set of hazards, many of which have the potential to escalate rapidly into large-scale emergencies. Effective preparation begins with an understanding of these hazards:

- **Blowouts**: One of the most serious and feared incidents in offshore drilling, a blowout occurs when the pressure within a reservoir forces oil, gas, or water to flow uncontrollably from the wellbore. Blowouts can lead to catastrophic fires, explosions, and significant oil spills, posing a severe threat to both personnel and the environment. The Deepwater Horizon disaster in 2010 remains a stark reminder of the potentially devastating impact of a blowout.

- **Oil Spills**: Oil spills can happen during the drilling, production, or transportation of oil and gas. Even a relatively small spill can have long-lasting effects on marine ecosystems, harming wildlife and damaging fisheries. Large-scale spills can result in enormous cleanup costs, damage to the reputation of the companies involved, and significant economic losses for coastal communities.

- **Fires and Explosions**: Offshore platforms house vast quantities of flammable materials, including oil, gas, and chemicals. The combination of these materials with potential ignition sources, such as electrical equipment, poses a constant risk of fire and explosion. Quick containment of any fire is critical to preventing a small incident from escalating into a full-scale disaster.

- **Equipment Failure**: Offshore operations depend on highly complex, interdependent systems and technologies, such as blowout preventers (BOPs), engines, and drilling rigs. A failure in any one of these systems can have cascading effects, leading to uncontrolled spills, explosions, or other dangerous situations.

- **Natural Hazards**: Offshore platforms and vessels are vulnerable to severe weather, including hurricanes, tropical storms, and rough seas. These natural events can disrupt operations, damage infrastructure, and pose serious risks to the safety of personnel. In rare cases, seismic activity, tsunamis, or underwater landslides can also threaten offshore installations.

1.2 Human Factors

While technological advancements have improved safety in offshore operations, human factors remain a critical element of risk. Many offshore emergencies are linked to human error, inadequate training, or poor decision-making in high-stress situations. Common human-related risks include:

- **Fatigue**: Offshore workers often endure long shifts in isolated environments, which can lead to fatigue. Fatigue impairs judgment, decision-making, and reaction times, increasing the likelihood of accidents.

- **Competence**: All personnel working offshore must be adequately trained to perform their duties, including responding to emergency situations. Insufficient training or lack of familiarity with safety procedures and equipment can exacerbate an emergency, increasing the risk to personnel and the environment.

1.3 Environmental and Economic Impacts

The consequences of an offshore emergency can be catastrophic, both environmentally and economically. Oil spills, for example, can devastate marine ecosystems, endanger wildlife, and contaminate fisheries, leading to long-term ecological damage. The economic costs of spills, fires, and other emergencies can also be substantial, including cleanup expenses, compensation claims, regulatory fines, and lost production. Public perception and company reputation are often severely impacted, which can have lasting effects on the organization's profitability and future operations.

Chapter 2

Regulatory Framework for Offshore Emergency Preparedness

2.1 International Standards and Guidelines

Offshore oil and gas operations are subject to stringent regulations and guidelines aimed at ensuring safety, environmental protection, and effective emergency preparedness. Several international bodies provide oversight and develop standards for the industry:

- **International Maritime Organization (IMO)**: The IMO plays a central role in regulating the shipping and offshore oil sectors. The organization establishes international guidelines for safety management, pollution prevention, and emergency preparedness through conventions such as MARPOL (Marine Pollution) and SOLAS (Safety of Life at Sea).

- **International Oil and Gas Producers (IOGP)**: The IOGP is a global organization that represents the upstream oil and gas industry. It publishes guidance on best practices for safety, emergency preparedness, environmental protection, and incident response.

- **International Safety Management (ISM) Code**: The ISM Code, developed by the IMO, mandates that offshore operators implement and maintain Safety Management Systems (SMS). These systems must include provisions for emergency preparedness, risk assessments, and response plans to ensure the safety of personnel, the environment, and assets.

2.2 National Regulations

Countries with offshore oil and gas operations have developed their own regulatory frameworks, which often build on international guidelines while addressing specific regional challenges. Two examples of national regulatory bodies are:

- **U.S. Bureau of Safety and Environmental Enforcement (BSEE)**: BSEE is responsible for enforcing safety and environmental regulations on the U.S. outer continental shelf.

The agency oversees offshore oil and gas operations, conducts inspections, and ensures that operators are adequately prepared to respond to emergencies.

- **European Offshore Safety Directive**: The European Union's Offshore Safety Directive sets out safety requirements for offshore operations in EU waters. It mandates that operators implement effective safety management systems and emergency response plans, and it establishes oversight bodies in member states to enforce these regulations.

2.3 Key Emergency Preparedness Requirements

International and national regulations typically require offshore operators to take several key steps to ensure emergency preparedness, including:

- **Risk Assessments**: Operators must conduct comprehensive risk assessments to identify potential hazards, evaluate the likelihood and severity of incidents, and implement mitigation measures.

- **Emergency Response Plans**: Detailed Emergency Response Plans (ERPs) must be developed, outlining the actions to be taken in the event of various emergency scenarios, including fires, explosions, oil spills, and blowouts.

- **Training and Drills**: Personnel must receive regular training on emergency procedures, including participation in safety drills that simulate real-life emergency scenarios. These drills help ensure that crew members are familiar with evacuation routes, firefighting equipment, and other response measures.

Chapter 3

Emergency Response Planning

3.1 Developing an Emergency Response Plan

An Emergency Response Plan (ERP) is a comprehensive document that outlines the procedures and actions to be taken in response to specific emergencies. A well-prepared ERP can save lives, protect the environment, and minimize financial losses in the event of a disaster. The development of an ERP involves several key steps:

- **Risk Assessment**: The first step in creating an ERP is to conduct a thorough risk assessment, identifying potential hazards such as equipment malfunctions, environmental threats, or human errors. This risk assessment serves as the foundation for determining the appropriate response measures.

- **Response Actions**: The ERP should detail the specific actions to be taken in the event of an emergency, including containment, evacuation, fire suppression, and oil spill response procedures. It must also address the steps required to protect personnel, mitigate damage to the environment, and restore normal operations.

3.2 Key Components of an Emergency Response Plan

Effective ERPs contain several key components:

- **Response Strategies**: This section outlines the strategies for mitigating the impact of emergencies. These strategies may include isolating affected areas, shutting down equipment, deploying firefighting systems, and activating oil spill containment measures.

- **Roles and Responsibilities**: Clearly defined roles and responsibilities ensure that all personnel understand their duties during an emergency. The ERP should designate an Incident Commander and outline the responsibilities of emergency teams, onshore responders, and external agencies.

- **Communication Protocols**: Communication is crucial during an emergency. The ERP must include protocols for both internal and external communication, ensuring that all relevant parties—such as crew members, regulatory bodies, and the media—are kept informed throughout the emergency.

3.3 Drills and Simulations

To maintain preparedness, offshore operators must conduct regular drills and simulations of emergency scenarios. These exercises allow personnel to practice their response actions, test the effectiveness of the ERP, and identify any weaknesses in the plan. Drills should cover a wide range of emergencies, including oil spills, fires, blowouts, and evacuations, to ensure a comprehensive response capability.

Chapter 4

Oil Spill Contingency Planning

4.1 What is Oil Spill Contingency Planning?

Oil spill contingency planning involves preparing for the possibility of an oil spill and developing strategies to respond quickly and effectively. Oil spills can cause widespread damage to marine ecosystems and coastal communities, making contingency planning a critical component of offshore emergency preparedness. A well-developed contingency plan outlines the actions to be taken in the event of a spill, helping to minimize environmental harm and reduce the financial costs of cleanup efforts.

4.2 Oil Spill Response Strategies

The response to an oil spill depends on several factors, including the size and type of spill, its location, and prevailing weather conditions. Common response strategies include:

- **Containment and Recovery**: Booms (floating barriers) and skimmers are deployed to contain the spilled oil on the surface of the water and recover as much oil as possible. This method is most effective in calm sea conditions.

- **Use of Dispersants**: Chemical dispersants can be sprayed on the oil to break it up into smaller droplets, allowing natural processes to degrade the oil more quickly. While dispersants can be effective, they may also have environmental side effects and must be used with caution.

- **In-Situ Burning**: In certain cases, controlled burning of the oil on the water's surface can be used to reduce the volume of the spill. However, this method produces air pollution and can only be used under specific conditions.

- **Shoreline Protection**: In the event of a large spill, protective measures such as booms and barriers are deployed to prevent oil from reaching sensitive coastal areas, including wetlands, beaches, and fisheries.

4.3 Role of Regional and International Cooperation

Responding to a large oil spill often requires international cooperation, particularly in cases where the spill threatens multiple countries. Regional response teams and international organizations such as the IMO facilitate collaboration, enabling countries to share resources, equipment, and expertise. This ensures that a well-coordinated and effective response can be mounted, even in cross-border emergencies.

Chapter 5: Offshore Evacuation and Rescue Operations

5.1 Evacuation Planning

Evacuation is one of the most critical aspects of offshore emergency response, particularly given the remote and isolated nature of offshore platforms. An effective evacuation plan must account for various scenarios, including fires, explosions, blowouts, and severe weather. Key components of an evacuation plan include:

- **Evacuation Routes and Procedures**: All offshore platforms must have clearly marked evacuation routes leading to lifeboats, rafts, or helicopter landing pads. These routes should be regularly tested to ensure accessibility in the event of an emergency.

- **Helicopter and Lifeboat Protocols**: Helicopters are often the primary means of evacuating personnel from offshore platforms. However, in cases where helicopters are unavailable or unsafe to use, lifeboats or life rafts must be fully operational and ready for deployment.

- **Muster Stations**: During an emergency, personnel are required to gather at designated muster stations, where they will be accounted for and provided with instructions for evacuation. Regular muster drills ensure that all crew members are familiar with their assigned stations and procedures.

5.2 Rescue Techniques

In some emergency situations, personnel may need to be rescued by external parties, such as emergency response teams from onshore or at sea. Rescue techniques include:

- **Helicopter Rescue Operations**: Helicopters can extract personnel from the platform or the water, depending on the nature of the emergency. Helicopter rescue operations require coordination between the offshore installation and rescue teams onshore.

- **Marine Vessel Assistance**: Rescue vessels can be deployed to the location to provide evacuation support, medical assistance, and firefighting capabilities.

- **Offshore Standby Vessels**: Many offshore platforms are required to have a standby vessel stationed nearby in case of an emergency. These vessels are equipped to perform rescue operations, provide first aid, and transport injured personnel to safety.

5.3 Challenges in Evacuation

Evacuating personnel from an offshore platform can be extremely challenging, particularly in severe weather conditions. High winds, rough seas, and communication failures can delay rescue efforts and complicate evacuation procedures. Emergency plans must account for these difficulties by including alternative evacuation routes and contingency measures for when primary evacuation methods are unavailable.

Chapter 6

Fire and Explosion Response in Offshore Operations

6.1 Understanding Fire and Explosion Hazards

Fires and explosions are among the most dangerous emergencies in offshore operations, given the presence of large quantities of flammable materials such as oil, gas, and chemicals. Fires can spread rapidly and cause extensive damage if not contained quickly. Explosions often result from gas leaks or the accumulation of flammable gases in confined spaces, and they can be triggered by electrical sparks or open flames.

6.2 Fire Suppression Systems

Offshore platforms are equipped with sophisticated fire detection and suppression systems to minimize the risk of fires and explosions. These systems include:

- **Automatic Fire Detection**: Sensors installed throughout the platform detect smoke, heat, or gas leaks and trigger alarms to alert personnel.

- **Deluge Systems**: Deluge systems are designed to release large volumes of water or foam to extinguish fires and prevent them from spreading. These systems are typically activated automatically when a fire is detected.

- **Portable Fire Extinguishers**: In addition to automated systems, crew members are trained to use portable fire extinguishers to control small fires before they escalate.

6.3 Firefighting Techniques and Equipment

All offshore personnel must be trained in firefighting techniques and familiar with the use of firefighting equipment. Key aspects of firefighting response include:

- **Firefighting Training**: Regular training ensures that personnel know how to operate firefighting equipment and extinguish different types of fires, including electrical fires and chemical fires.

- **Use of Breathing Apparatus**: In the event of a fire, crew members may need to use breathing apparatus to protect themselves from inhaling toxic fumes.

- **Coordination with Onshore Response**: In severe cases, firefighting efforts may require assistance from onshore emergency response teams, who can provide additional resources and expertise.

Chapter 7

Managing Human Factors in Emergencies

7.1 Psychological Response to Emergencies

Emergencies in offshore environments can trigger intense psychological reactions among personnel, such as fear, anxiety, and panic. These responses can hinder effective decision-making and lead to mistakes during critical moments. To manage human factors during emergencies:

- **Stress and Decision-Making**: High-stress situations can impair judgment, making it difficult for personnel to make clear and informed decisions. Regular training and emergency drills help individuals remain calm and focused during emergencies.

- **Leadership**: Strong leadership is essential in managing human factors. The Incident Commander plays a key role in providing clear instructions, maintaining control, and reassuring personnel to prevent panic.

7.2 Human Reliability

Human reliability refers to the ability of personnel to perform their tasks correctly, even under extreme pressure. Several factors contribute to human reliability, including:

- **Training**: Regular training ensures that all personnel understand their roles and responsibilities in an emergency, reducing the likelihood of mistakes.

- **Competence**: Each crew member must be competent in their specific duties, from operating machinery to performing medical tasks or handling firefighting equipment.

7.3 Fatigue Management

Fatigue is a significant risk factor in offshore operations, where long shifts and extended periods away from shore are common. Fatigue can slow reaction times and impair judgment during emergencies. Effective fatigue management strategies—such as rotating shifts, ensuring adequate rest periods, and monitoring crew well-being—are critical to maintaining alertness and reducing the risk of errors.

Conclusion

The offshore oil and gas industry operates in some of the most challenging environments on Earth, where emergencies can quickly escalate and have catastrophic consequences. Effective emergency preparedness and response are critical to ensuring the safety of personnel, protecting the environment, and minimizing financial losses. By developing comprehensive emergency response plans, conducting regular drills, and ensuring that personnel are well-trained and equipped, the offshore industry can mitigate the risks associated with operations and safeguard its assets.

Emergency preparedness is not merely a regulatory requirement but a fundamental aspect of operational success. In an industry where lives, resources, and environmental health are at stake, the ability to respond swiftly and effectively to emergencies is essential. The lessons learned from past disasters, combined with continuous improvements in technology, training, and safety management, will help the industry maintain high standards of safety and preparedness in the future.

Glossary: Emergency Preparedness and Response for Offshore Oil & Gas

1. **Automatic Fire Detection**: Sensors used to detect smoke, heat, or gas leaks and trigger alarms on an offshore platform to prevent fires from escalating.

2. **Blowout**: The uncontrolled release of oil, gas, or fluids from a well due to pressure control failure, leading to potential fires, explosions, or spills.

3. **Blowout Preventer (BOP)**: A device installed at the wellhead to control well pressure and prevent blowouts during drilling operations.

4. **Boom**: A floating barrier used to contain oil spills on the water's surface, preventing the spread of oil to unaffected areas.

5. **Containment Boom**: A physical barrier designed to prevent the spread of oil during an oil spill response, facilitating recovery efforts.

6. **Deluge System**: A fire suppression system that releases large volumes of water or foam to extinguish fires on an offshore platform.

7. **Dispersants**: Chemicals applied to oil spills to break up slicks into smaller droplets, making it easier for natural degradation processes to occur.

8. **Drillship**: A specialized vessel equipped with drilling equipment used for offshore oil and gas exploration and production activities.

9. **Emergency Evacuation Procedures**: Guidelines and protocols that detail how personnel should evacuate an offshore platform safely during emergencies.

10. **Emergency Response Plan (ERP)**: A comprehensive document that outlines procedures for responding to various offshore emergencies, including oil spills and blowouts.

11. **Environmental Impact Assessment (EIA)**: A process to evaluate the potential environmental effects of offshore oil and gas activities on marine ecosystems and wildlife.

12. **Evacuation Plan**: A strategic plan detailing how personnel will evacuate an offshore platform during emergencies, including routes and procedures for lifeboats and helicopters.

13. **Fatigue Management**: Strategies used to minimize the impact of fatigue on offshore personnel, reducing human errors during emergencies.

14. **Fire and Explosion Hazards**: The risks associated with fires and explosions on offshore platforms due to the presence of flammable materials like oil and gas.

15. **Fire and Gas Detection System**: Equipment used to detect fires and gas leaks on an offshore platform and initiate emergency responses.

16. **Fire Suppression System**: Equipment installed on offshore platforms to extinguish fires and prevent their spread, including sprinklers, foam systems, and deluge systems.

17. **First Response Team (FRT)**: The group responsible for initiating immediate actions during an offshore emergency, providing support until external assistance arrives.

18. **Helideck**: A designated helicopter landing pad on an offshore platform used for transporting personnel and emergency evacuations.

19. **Human Factors**: The influence of human behavior, such as decision-making and fatigue, on the safety and effectiveness of offshore emergency responses.

20. **Incident Commander**: The individual responsible for managing and leading the emergency response during an offshore incident.

21. **In-Situ Burning**: A technique used to manage oil spills by igniting the oil on the water's surface to reduce the volume of spilled material.

22. **International Maritime Organization (IMO)**: A United Nations agency responsible for developing international regulations for maritime safety and the prevention of marine pollution.

23. **International Oil and Gas Producers (IOGP)**: An industry organization that promotes best practices in safety, emergency preparedness, and environmental protection for offshore operations.

24. **International Safety Management (ISM) Code**: A framework for implementing safety management systems in offshore operations to ensure safety and emergency preparedness.

25. **Lifeboat Launching Protocol**: Procedures for the safe and efficient deployment of lifeboats during an offshore emergency evacuation.

26. **Marine Pollution (MARPOL)**: An international convention designed to prevent pollution from ships and offshore platforms, including oil spills and hazardous substances.

27. **Master Response Plan (MRP)**: A comprehensive emergency plan that encompasses all aspects of offshore emergency preparedness, including evacuation, fire suppression, and oil spill response.

28. **Muster Station**: A designated location on an offshore platform where personnel gather during an emergency to receive evacuation instructions and accountability.

29. **National Oil Spill Contingency Plan (NOSCP)**: A government-regulated framework that outlines how national agencies and operators will respond to oil spills.

30. **Oil Spill Contingency Plan (OSCP)**: A plan developed by offshore operators to respond effectively to oil spills, including containment, recovery, and cleanup measures.

31. **Personal Protective Equipment (PPE)**: Safety gear, such as helmets, gloves, and life jackets, used by offshore personnel to protect against hazards during emergencies.

32. **Platform Supply Vessel (PSV)**: A ship used to transport supplies, equipment, and personnel to offshore platforms, playing a key role in logistics and emergency response.

33. **Port State Control (PSC)**: The inspection of foreign ships by national authorities to ensure compliance with international safety and environmental regulations.

34. **Preventive Maintenance**: Regular maintenance conducted on offshore equipment to reduce the risk of failure and potential emergencies.

35. **Risk Assessment**: The process of identifying, analyzing, and mitigating potential hazards on offshore platforms to prevent incidents.

36. **Safety Management System (SMS)**: A structured approach to managing safety in offshore operations, including emergency preparedness and regulatory compliance.

37. **Search and Rescue (SAR)**: Coordinated operations that involve locating and rescuing personnel in distress or missing at sea, often using helicopters and vessels.

38. **Shut-in**: The process of closing down a well or ceasing production during an emergency, such as a blowout or equipment failure, to prevent further escalation.

39. **Slick**: A thin layer of oil spread on the water's surface, typically resulting from an oil spill.

40. **Spill Response Team (SRT)**: A team of trained personnel responsible for managing and responding to oil spills in offshore environments.

41. **Standby Vessel**: A ship stationed near offshore platforms, equipped to perform rescue operations, provide medical assistance, and support evacuations during emergencies.

42. **Steam Methane Reforming (SMR)**: A method used to produce hydrogen, often utilized in offshore operations as part of energy and decarbonization efforts.

43. **Subsea Blowout**: A blowout occurring beneath the water's surface, posing additional challenges for containment and recovery operations.

44. **Tropical Cyclone**: A powerful storm system that can threaten offshore platforms, leading to damage and the need for emergency evacuations.

45. **Tsunami Evacuation Plan**: A plan designed to safely evacuate personnel from offshore platforms in the event of a tsunami or other marine-related natural disasters.

46. **Unified Command**: A leadership structure used during emergency responses to bring together multiple agencies and stakeholders for coordinated action.

47. **Upstream**: The sector of the oil and gas industry involved in exploration, drilling, and production, including offshore operations.

48. **Well Control**: Techniques used to maintain control over a well during drilling operations to prevent blowouts and other emergencies.

49. **Wellhead**: The structure located at the surface of an offshore well that controls the flow of oil, gas, or other fluids.

50. **Wild Well**: A well that is out of control due to a blowout or other incident, requiring specialized intervention to regain control and prevent environmental damage.

www.ingramcontent.com/pod-product-compliance
Lightning Source LLC
Chambersburg PA
CBHW030041230526
45472CB00002B/623